Mach 3 and Beyond:

The Untold Story of SR-71 #61-7971

By

C.S. Duncan

Chapter 1: "The Skunk Works Miracle"

The year was 1965, and I was born in a place known for making the impossible possible. Lockheed's Skunk Works was a hive of innovation, a place where the brightest minds in aviation came together to push the boundaries of what was known. I could feel it in the air as the engineers worked day and night, my titanium skin taking shape piece by piece. They crafted me to fly and conquer the sky, designed to touch the edge of space and do so faster than anything that had ever flown before me.

I was built to withstand speeds that would melt lesser aircraft. My frame, forged from titanium and other heat-resistant materials, would expand during flight as I sped through the atmosphere. The men who built me knew this. They crafted me with expansion joints, allowing me to stretch and shift as I raced past Mach 3. Every panel was meticulously placed, every part tested to endure the extreme forces I would face. I was a product of genius, and I was ready to prove it.

November 17, 1966—the day I first felt the air beneath my wings. As I lifted off the runway for the first time, a surge of power coursed through my engines, and I knew I was destined for something great. The roar of my Pratt & Whitney J58 engines echoed in my ears, and the sky opened up before me. I climbed higher and higher, pushing through the clouds, my titanium skin shimmering in the sunlight. At 85,000 feet, I could see the curvature of the Earth, the blue sky fading into the blackness of space. I was not just an aircraft. I was a pioneer.

Chapter 2: "Into the Fray"

Assigned to the 9th Strategic Reconnaissance Wing at Beale Air Force Base, California, my missions began almost immediately. The Cold War was at its height, and the world was a dangerous place filled with secrets and shadowed enemies. My purpose was clear: to gather intelligence from altitudes and speeds that made me untouchable. I was the Blackbird, a name that suited me well, and I was built to fly into danger and come out unscathed.

The first missions were exhilarating. I would climb to cruising altitude, my engines howling as I reached Mach 3, flying faster than any aircraft before me. The ground beneath me blurred, and the sky above was my domain. I was tasked with reconnaissance over hostile territories, my cameras capturing every detail from miles above. Vietnam, Eastern Europe, the heart of the Cold War—I flew over them all, silent and unseen.

Missiles were fired at me, but they could never reach me. I could sense them coming, my advanced warning systems alerting my pilot of the threat. And then, we would accelerate. There was nothing on Earth that could match my speed. I could hear the tension in the cockpit when the alarms blared, but the calm certainty of my pilot always followed it. He knew, as I did, that no missile could catch me. We would laugh in the face of danger, a bond forged between man and machine.

Chapter 3: "The Edge of Speed"

Every mission I flew, I felt the world watching, though no one could see me. I was a shadow at the sky's edge, moving too fast for radar to track. I flew higher than any aircraft, faster than the fastest bullet. At Mach 3, I could cover vast distances in minutes. It was an exhilarating dance with the very limits of physics. My sleek, black frame was designed for this—optimized to reduce my radar signature, I was practically invisible. The enemy could only guess where I had been after I had already left.

The pilots who flew me were a special breed, trained to withstand the intense physical toll of flying at such speeds and altitudes. They wore pressure suits, their helmets connected to oxygen supplies, and their bodies braced for the high-G forces. I could feel their tension and excitement every time they climbed into my cockpit. Together, we would become one—pilot and plane—flying through the sky as if we owned it. In many ways, we did.

There was an unspoken trust between us. My pilot relied on me to keep him safe, and I relied on him to push me to my limits. Each mission tested our abilities, a challenge to fly farther, faster, and higher. And each time, we succeeded. We were unstoppable.

Chapter 4: "Giant Reach"

The world was on edge in the early 1970s. Tensions in the Middle East rose, and war loomed on the horizon. In January 1974, I was called upon for a mission that would define me: Operation Giant Reach. It was a long-range reconnaissance mission that would take me over the Middle East during the Yom Kippur War, gathering intelligence crucial to the balance of power in the region.

My body hummed with energy as my engines powered up. My mission: to fly from Seymour-Johnson Air Force Base in North Carolina to the heart of the conflict and back in just over 10 hours. It was a test of endurance, of speed, and precision. I was ready.

As I streaked over the desert at Mach 3, I felt the heat of the land below and the tension in the air as battles raged on the ground. But up here, I was untouchable. My cameras clicked away, capturing the details of military positions, troop movements, and strategic points. The intelligence I gathered would help Israel counter the forces against them, a silent but vital role in the unfolding drama.

After more than 10 hours in the air, I returned home, and my mission was successful. I had flown farther and faster than I had ever done before, proving that no distance was too great for me. I was a machine of war but also a guardian of peace. I had fulfilled my purpose.

Chapter 5: "The Silent Years"

By 1990, the world had changed. The Cold War was fading, with it, the need for reconnaissance flights like mine. The Air Force decided to retire me, and for the first time, I found myself grounded. It was a strange feeling to be still after so many years in the sky. I was no longer needed, no longer the fastest, the highest, the untouchable.

But my story wasn't over. NASA came calling, offering me a new lease on life. I was no longer a weapon but a tool of science. I joined their fleet, helping them push the boundaries of high-speed flight research. It wasn't the same thrill as reconnaissance missions, but it was a new kind of purpose. I flew to help understand the limits of aeronautical engineering and to test the boundaries of human knowledge. It was quieter, more reflective work, but I learned to appreciate the calm after years of high-stakes missions.

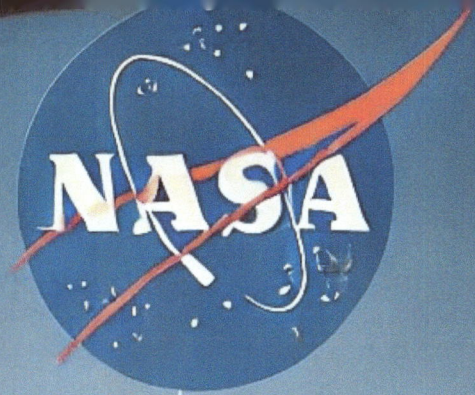

NASA fitted me with new sensors and research equipment, transforming me into a flying laboratory. My once sleek, black frame now bore markings of a different kind—NASA's insignia, signifying my new role. I no longer carried the weight of national security on my wings; instead, I brought the hopes of scientists eager to unlock the secrets of high-speed, high-altitude flight.

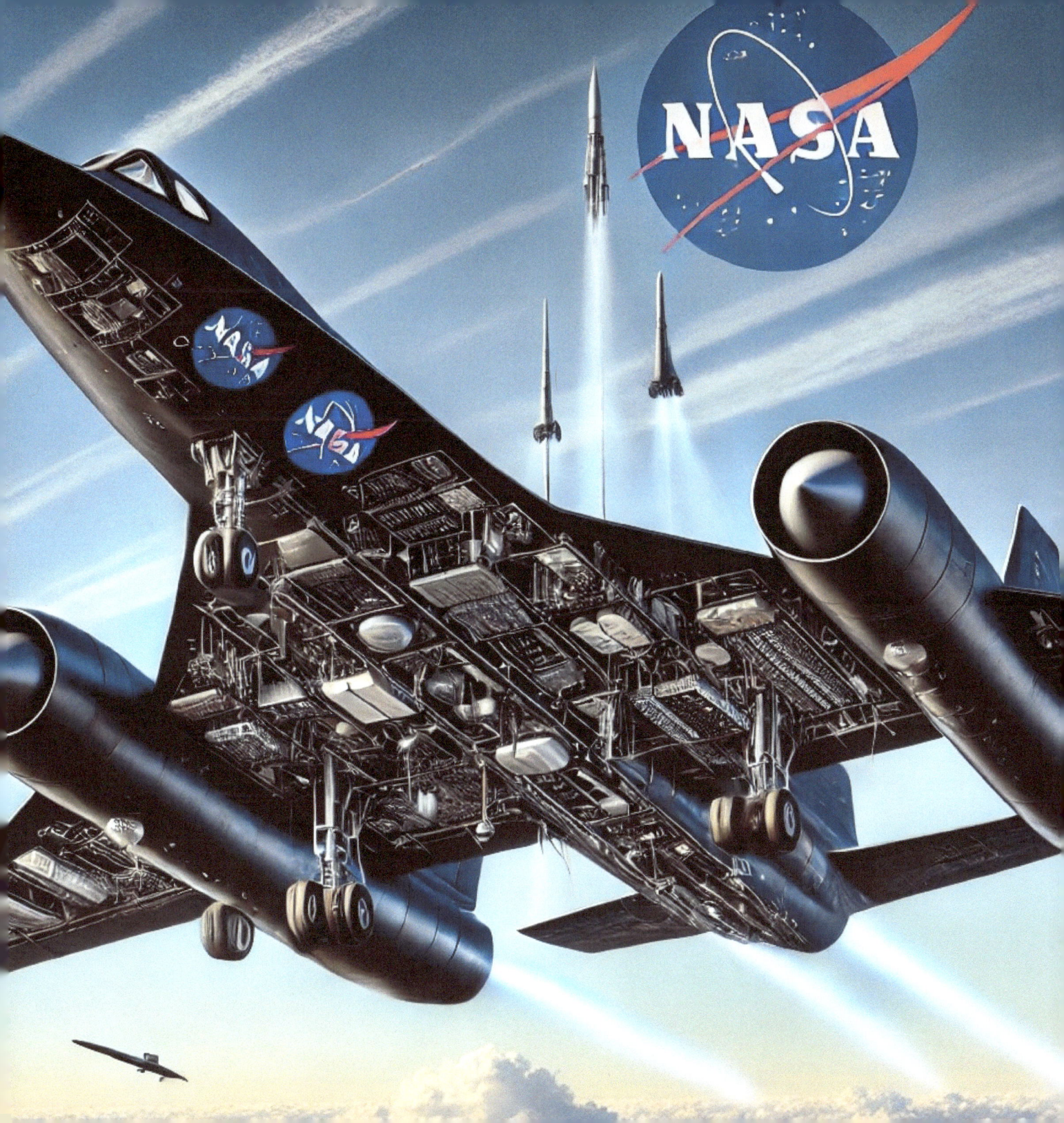

Still, part of me missed the excitement of my former life. The skies called to me, and I longed to return to the missions that had once defined me.

Chapter 7: "Reaching for the Stars"

With NASA, I participated in projects that pushed the boundaries of human knowledge. My flights were no longer about outpacing missiles or photographing enemy installations. They were about understanding the very nature of speed, aerodynamics, and the atmosphere. I helped researchers study high-altitude atmospheric conditions in one project, flying higher than most aircraft could ever dream.

The scientists at NASA were fascinated by my ability to handle the extreme conditions of flight at such incredible speeds. My titanium frame, which once expanded and contracted with the heat of Mach 3, was now the subject of intense study. They wanted to know how I withstood such punishment and how my design could inform the next generation of aircraft.

I flew missions to gather data on atmospheric conditions, contributing to research that would later influence spacecraft design and even the development of commercial supersonic jets. It was a quieter, more measured pace than the urgent reconnaissance flights of my past. But I found peace in it, knowing I was still pushing the boundaries of what was possible.

Chapter 8: "A New Kind of Legacy"

During my time with NASA, I began to realize that my legacy would not only be tied to the secrets I gathered during the Cold War. I had become something more than a spy plane—I had become a symbol of human ingenuity. The scientists I worked with treated me with the same respect that my Air Force pilots did, but their goals were different. They were not interested in enemy movements or missile installations. They were interested in the future.

As I flew high over the desert, gathering data for NASA's research projects, I began to reflect on my journey. From my early days as a classified project at the Skunk Works to my missions over Vietnam and the Middle East, I had served with honor. But now, I was part of something bigger than just military operations. I was contributing to knowledge that could shape the future of flight itself.

I thought back to the pilots who had flown me, to the engineers who had built me, and to the scientists who now studied me. We were all part of the same journey— a journey to understand what it meant to fly higher, faster, and farther than ever before. I was proud of my role, and I knew that my story was far from over.

Chapter 9: "The Final Flight"

In 1997, the Air Force called me back for one last mission. It was unexpected, but I welcomed the challenge. I had been with NASA for nearly a decade, but my heart still longed for the thrill of military service. I was flown to Edwards Air Force Base, where I was prepped for my final flight.

The mission was simple—a test of my systems, a final demonstration of my capabilities before I retired. As I roared into the sky one last time, I felt a rush of emotions. This was my final journey, and I wanted to savor every moment. The sun glinted off my titanium skin as I climbed to 85,000 feet, the familiar sensation of speed enveloping me.

I flew faster than I ever had, my engines roaring as I cut through the thin atmosphere. My pilot was calm and focused, but I could sense the emotion in the cockpit. We both knew that this was the end of an era. The world had changed, and my time was running out.

Satellites were replacing me, offering a cheaper, more permanent solution to reconnaissance. We accepted this reality, knowing I had done everything I was built for. As I descended, my mission complete, I felt a sense of peace. I had done everything I was built for and more.

When I touched down at Edwards Air Force Base, I knew my flying days were over. But I also knew that I had left an indelible mark on the history of aviation. I had been the fastest, the highest, the untouchable. Now, it was time to rest.

Chapter 10: "A Museum Piece"

2002, I found myself at the Evergreen Aviation & Space Museum in McMinnville, Oregon. My flying days were behind me, but my story was far from over. I had become a symbol of human achievement, a relic of a time when the sky was the final frontier. People came from all over to see me, to marvel at the sleek, black design that had once outrun missiles and gathered intelligence from the stratosphere.

I watched as children pressed their faces against the glass, their eyes wide with wonder. They didn't know the details of my missions, but they knew that I had been something special. Veterans, too, came to see me, their eyes filled with memories of when I ruled the skies.

I may no longer fly, but I continue to inspire. My body may be grounded, but my legacy soars. I am a reminder of what we can achieve when we dare to dream and push the boundaries of what is possible. My story is about speed, power, and the human spirit. And even in stillness, I am alive. I am SR-71 #61-7971, the Blackbird, and I was born to go Mach 3 and beyond.

Facts you might not know about the SR-71 Blackbird:

1. Fuel Leaks on the Ground:

The SR-71 was designed with loose-fitting titanium panels that expanded at high speeds due to heat, sealing the aircraft during flight. This caused it to leak fuel (JP-7) when on the ground.

2. Outran Missiles:

Despite being targeted by over 4,000 missiles during its service, none ever hit the SR-71. The aircraft's speed and altitude, combined with its ability to jam radar, made it virtually untouchable.

3. Black Paint for Stealth and Heat Management:

The black paint was not just for aesthetics—it contained ferrite particles to absorb radar waves and dissipate heat, playing a role in stealth and temperature regulation.

4. Fuel as a Coolant:

The SR-71's JP-7 fuel served double duty, acting as a coolant to manage the extreme temperatures generated during flight at Mach 3+ speeds.

5. Used for NASA Research:

NASA used the SR-71 for high-speed and high-altitude aeronautical research. Its missions included tracking celestial objects and studying the ozone layer.

6. Expands in Flight:

Due to the intense heat generated at Mach 3+, the SR-71 expanded in length by several inches while in flight, making the use of expansion joints essential in its design.

7. Titanium from the Soviet Union:

The U.S. had to acquire much of the titanium used in the SR-71's construction from the Soviet Union via third parties, as it was the only metal that could withstand the heat at such high speeds.

8. Never Needed Defensive Armament:

The SR-71 carried no weapons. Its defense was purely its speed and altitude—outrunning threats was its primary means of protection.

9. Pilots Wore Space Suits:

Due to the extreme altitudes the SR-71 operated at (above 85,000 feet), pilots had to wear pressurized suits similar to astronaut gear to protect against cabin depressurization and extreme conditions.

10. Shock Diamonds in Exhaust:

The SR-71's continuous afterburning engines created visible shock waves in the exhaust, known as "shock diamonds," a result of its unique thrust production.

11. Specialized Tires:

The SR-71's tires were filled with nitrogen and reinforced with aluminum to withstand the high pressures (up to 415 psi) and heat generated during takeoff and landing.

12. JP-7 Fuel's High Flashpoint:

The JP-7 fuel had such a high flashpoint that it couldn't be ignited with a standard flame, making it much safer to handle despite the aircraft's high speeds and fuel leaks.

13. Set World Records:

The SR-71 still holds the official records for the fastest air-breathing manned aircraft and the highest sustained altitude flight, set in 1976.

14. Used to Test Satellite Communication:

The SR-71 played a key role in testing Motorola's Iridium Satellite communication system, essentially acting as a moving satellite to test ground transmitters and receivers.

15. Expansion of Pilot Physiology:

Pilots had to undergo astronaut-like physical training and conditioning due to the high G-forces and extreme conditions they encountered during flights.

16. Special Fuel Ignition:

Triethylborane (TEB) was used to ignite the SR-71's engines, as regular spark plugs couldn't handle the high compression and conditions within the engine.

17. The Radar Cross-Section Was Tiny:

The SR-71's radar cross-section was small for such a large aircraft. It used stealth design elements, including angled surfaces and radar-absorbing materials, long before modern stealth aircraft were developed.

18. High Operational Costs Led to Retirement:

Despite its performance, the high cost of maintaining and operating the SR-71 was a significant factor in its retirement, especially as satellite reconnaissance improved.

19. No Significant Airframe Fatigue:

The SR-71's titanium construction held up so well under repeated high-speed flights that it showed little airframe fatigue, even after thousands of hours of service.

20. Final Flight Was a Record Breaker:

The SR-71's final flight on March 6, 1990, set a coast-to-coast speed record, flying from Los Angeles to Washington, D.C. in just 64 minutes and 20 seconds at an average speed of 2,144 mph.